KYLE LASCURETTES

The Concert of Europe and Great-Power Governance Today

What Can the Order of 19th-Century Europe Teach Policymakers About International Order in the 21st Century?

Prepared for the Office of Net Assessment,
Office of the Secretary of Defense
Approved for public release; distribution unlimited

Contents

What Was the Concert of Europe?..2

What Were the Concert's Foundational Principles?...5

Why Was the Concert Considered Desirable?..8

When and Why Did the Concert Decline?...14

What Can We Learn from the Concert?...17

Appendix..23

Notes..26

Bibliography..30

About the Author...33

The RAND Corporation is a research organization that develops solutions to public policy challenges to help make communities throughout the world safer and more secure, healthier and more prosperous. RAND is nonprofit, nonpartisan, and committed to the public interest.

Cover: Library of Congress map from 1854.

Limited Print and Electronic Distribution Rights

This document and trademark(s) contained herein are protected by law. This representation of RAND intellectual property is provided for noncommercial use only. Unauthorized posting of this publication online is prohibited. Permission is given to duplicate this document for personal use only, as long as it is unaltered and complete. Permission is required from RAND to reproduce, or reuse in another form, any of our research documents for commercial use. For information on reprint and linking permissions, please visit www.rand.org/pubs/permissions.html.

RAND's publications do not necessarily reflect the opinions of its research clients and sponsors. **RAND**® is a registered trademark.

For more information on this publication, visit www.rand.org/t/PE226.

© Copyright 2017 RAND Corporation

> The French Revolution had dealt a perhaps mortal blow to the divine right of kings; yet the representatives of this very doctrine were called upon to end the generation of bloodshed. In these circumstances, what is surprising is not how imperfect was the settlement that emerged, but how sane; not how "reactionary" according to the self-righteous doctrines of nineteenth-century historiography, but how balanced. It may not have fulfilled all the hopes of an idealistic generation, but it gave this generation something perhaps more precious: a period of stability which permitted their hopes to be realized without a major war or a permanent revolution.
>
> —Henry A. Kissinger[1]

A young Henry Kissinger wrote these words years before his own extraordinary tenure at the helm of American foreign policy. Yet in the decades since, Kissinger has only continued to heap praise on the founders of the so-called Concert of Europe.[2] Moreover, he has not been the only prominent American policymaker to be influenced by the era: Woodrow Wilson consciously envisaged the Versailles settlement and League of Nations Covenant to be the antithesis of the Concert of Europe's "Vienna System," but after the League of Nations failed to avert World War II, Franklin Roosevelt used the Concert as a guidepost for designing the great-power consortium that would become the United Nations Security Council (UNSC).[3] As historian Mark Mazower recently observed, Kissinger and Roosevelt were "not alone in looking to the past to help guide the world toward a less chaotic future, nor in finding in the long peace of the nineteenth century a golden age of farsighted statecraft."[4] Indeed, the observation that American policymakers today should adopt the grand strategic playbook of the European Concert's architects has almost become a truism.[5]

Yet for all of the contemporary attention and admiration it engenders, the European Concert of the 19th century continues to bedevil those seeking to unpack its intricacies and precise mechanisms. Even after all of this time, scholarship remains divided on nearly every fundamental aspect of the Concert of Europe: what it was, how long it lasted, what factors were key to its early successes and later decline, and what lessons, if any, policymakers should take from the Concert today. This paper takes up these issues in turn. Specifically, it is organized by five central questions: What precisely was the Concert of Europe? What were its core, foundational principles? What demonstrable effects did the Concert have on state behavior and international outcomes? When did the Concert system fall apart, and what were the causal roots of its demise? And finally, what lessons might American policymakers take from the European Concert experience as they attempt to advance their own vision for international order today?

To answer these questions, I have surveyed the comprehensive secondary historical canon on the Concert of Europe, mostly the work of diplomatic historians and political scientists. Where interpretations are corroborated across this literature with little or no significant dissent, I treat them as scholarly consensus. Where interpretations are in conflict in this secondary literature, I go to primary sources—the original texts of the relevant treaties or the correspondence of the relevant policymakers—to adjudicate which of the competing interpretations enjoys the greatest support.

What Was the Concert of Europe?

Concert of Europe scholarship—in both the historical and international relations academic disciplines—can be broadly divided into three camps. Each of these groupings corresponds with one of the three most-prominent paradigms in international relations theory: *realists*, who emphasize the primacy of international power distributions and states' national self-interests; *liberals*, who focus on the abilities of international organizations to mitigate the effects of international anarchy emphasized by realists and help even selfish states achieve common goals; and *constructivists*, who point to the importance of transnational collective identities that purportedly form between states as a product of their interactions. Briefly revisiting each of these interpretations of the Concert is instructive. I argue that the Concert of Europe most closely approximated an amalgamation of the liberal and constructivist perspectives.

When it comes to the European system of the 19th century, realists are the Concert skeptics, arguing either that the Concert never actually existed or that it consisted only of high-minded rhetoric that had little tangible effect on state behavior.[6] Even though scholars who have adopted a realist perspective on the Concert era—such as Korina Kagan, Matthew Rendall, and Branislav Slantchev—have usefully highlighted some of the shortcomings in Concert-related scholarship, they have done less well in combatting the substantial evidence that leaders spoke much differently and states behaved much differently than realist theory would expect throughout much of the 19th century.[7] Moreover, the realist claim that the European Concert rested on little more than traditional "balance-of-power" thinking—a grand strategic maxim that had already dominated European diplomacy for at least 100 years prior to the Concert—is a position that is no longer regarded as tenable in the vast majority of Concert scholarship. As Jennifer Mitzen recently put it, the realist perspective

> overlooks the fact that the leaders themselves intended to manage Europe together. They did not think that their redrawing of the map of Europe and redistribution of territory and political control . . . had the force to keep the peace. That is precisely why they chose to renew the alliance in November 1815 and insert an article in that treaty that called for consultation.[8]

For all their ongoing disagreements over the Concert, diverse scholars have generally agreed that its architects viewed what they were doing as, in many ways, a repudiation of balance-of-power diplomacy, not a reaffirmation of it.[9]

Scholars of the liberal international relations persuasion, by contrast, treat the Concert as a primitive but successful version of a conflict-mediating international organization.

Specifically, liberals characterize the Concert as either (1) a multilateral forum that decreased the difficulties of negotiation among many parties (transaction costs) while increasing the reliable information each could learn about one another in these negotiations (transparency) or (2) a particularly institutionalized version of a military alliance that contained novel provisions for limiting the exercise of raw power by its most-dominant members, Great Britain and Russia.[10]

Finally, international relations constructivists focus less on these formal institutional processes and instead characterize the Concert as a successful instance of normative convergence and consensus among the great powers of Europe that, depending on the particular account, developed collective intentions and interests or even a common transnational identity.[11]

While there are merits to each of these perspectives, both the liberal and constructivist interpretations also have their limitations. Against the liberals, characterizing the Concert as even a primitive international organization comparable to the United Nations or the North Atlantic Treaty Organization (NATO) seems problematic. The Concert had no formal or permanent bureaucracy and—outside of the initial treaties and whatever notes, minutes, or declarations individual delegations decided to write down—codified virtually nothing about its procedures, processes, and principles. In fact, a number of scholars attribute the Concert's effectiveness not to robust collective security guarantees or institutionalized commitments but to a distinct *lack* of these features that allowed it to remain flexible and adaptive to changing circumstances.[12] Yet against the constructivist view, socialization in the Concert was nowhere near as advanced as it is today in, say, the European Union, evidenced in part by states' frequent use of coercive threats against one another in Concert meetings (even as they only rarely acted on them).[13] Furthermore, any normative consensus that did take place was almost entirely limited to a small group of elites, not their national publics. Notable constructivist accounts even admit that the lack of deeper-level socialization might have eventually led to the Concert's demise.[14]

This paper thus adopts a conceptualization of the Concert between the liberal and constructivist views. At its core, the Concert of Europe was an agreement among the elite statesmen of Europe's great powers to adhere to and enforce a particular set of principles in their relations with one another on the European continent. Sometimes these principles seemed to exist as informal but powerful

> At its core, the Concert of Europe was an agreement among the elite statesmen of Europe's great powers to adhere to and enforce a particular set of principles in their relations with one another on the European continent.

"norms" in the constructivist sense that their normative content was deeply internalized by the actors that shared them. At other times, these rules had their greatest impact as instrumental focal points that provided self-interested and disagreeable countries a set of procedures to more easily avert wasteful, suboptimal conflict, a perspective more in line with the liberal view. Yet regardless of whether or not they codified into international law or represented deeply held values and identities shared by the leaders who advocated them, what mattered most was that these principles were both widely recognized (in rhetoric) and widely practiced (in behavior) by the great powers throughout the Concert era.

The Concert of Europe was thus a collection of often informal but nonetheless influential rules agreed to by the great powers of Europe—Great Britain, Austria, Russia, Prussia, and, later, France—after the French Revolutionary and Napoleonic Wars.[15] Great Britain was the dominant power in terms of wealth and actualized resources, while Russia was destined to play a large role in the post-Napoleonic system as the polity with the most potential power. Although Austria could not rival either of these states in material power, its centrality to the fate of the German states in the heart of Europe, combined with the famous diplomatic tact of its foreign minister, Prince Klemens von Metternich, assured it a first-tier position at the peacemaking table. Along with Metternich, the Concert's principal architects included Great Britain's powerful foreign secretary, Viscount Castlereagh, and the eccentric and unpredictable tsar of Russia himself, Alexander I. The chief minister of Prussia, Prince Karl August von Hardenburg, played an important but comparably lesser role in negotiating the final settlements, as did the foreign minister to the newly restored Bourbon Monarchy in France, Prince Charles Maurice de Talleyrand.

In crafting a settlement following the French Revolutionary and Napoleonic Wars, these leaders sought to address and forestall two interrelated concerns for the future: (1) subsequent bids for continental hegemony that could be as destructive and nearly as successful as Napoleon's had been and (2) radical, revolutionary movements that might externalize aggression and trigger system-wide regime change, as France's had during the Revolutionary and Napoleonic Wars. In reality, the first was merely the latest manifestation of an old and recurring fear of a single power dominating all of Europe, addressed previously in the major peace settlements of Westphalia in 1648 and Utrecht in 1713. Yet that fear had been given new life by the more novel and potent threat of revolutionary nationalism that had transformed revolutionary France into a menace that had almost achieved total hegemony over Europe.[16]

The leaders who first gathered in 1814 to forge a settlement that would end those wars were resolved to prevent anything like this from happening again. Collectively, the foundational principles produced in these negotiations are known as the "Vienna System," a reference to the famous Congress of Vienna of 1813–1814. That months-long meeting of the major foreign policy representatives of Europe's polities produced only a small portion of the rules that came to define the Vienna System. But it fostered much of the *spirit* that led to the remarkable instances of cooperation in the decades thereafter.[17] The major landmarks of the Vienna System and their significance for the emergent order now referred to as the Concert of Europe are summarized in Table 1.

TABLE 1
Core Agreements of the Vienna System

DATE	AGREEMENT	SIGNIFICANCE FOR THE EUROPEAN ORDER
March 1814	Treaty of Chaumont	The allies in war against France pledge to negotiate with France to end the war only as a collective, not separately; they also pledge to keep the alliance together for an unprecedented length of at least 20 years.
May 1814	First Treaty of Paris	In a separate, secret clause to this treaty that dictated the terms of peace with France, the great powers—Great Britain, Austria, Russia, and Prussia—grant themselves status and rights as "great powers" to dictate and enforce the terms of peace on the continent.
June 1815	Vienna Final Act	Separate negotiations and treaties at the Congress of Vienna produce a summary and collection of the massive territorial settlements; this is significant for implying that these separate settlements are now part of a united whole and are not reducible to their individual components.
September 1815	Holy Alliance	Austria, Russia, and Prussia sign a vague agreement to come to each other's aid; today, this is largely viewed as an empty agreement that was only later used by the eastern powers to justify anti-liberal interventions across the continent.
November 1815	Second Treaty of Paris	This treaty produces a slightly more punitive settlement against France in the wake of Napoleon's escape and defeat at Waterloo; it contains the most concrete language to date implying that anti-liberal/anti-revolutionary sentiment is a significant component of the emerging postwar order.
November 1815	Treaty of Alliance (Quadruple Alliance)	This treaty renews the long-term and general alliance made in the Treaty of Chaumont; it declares periodic meetings for the great powers to maintain peace and tranquility together throughout Europe.
October 1818	Treaty of Aix-la-Chapelle	France is welcomed into the Quadruple (now Quintuple) Alliance, Concert system, and great-power partnership; this treaty more strongly reiterates the great powers' agreement to hold periodic meetings to address European geopolitical matters.

What Were the Concert's Foundational Principles?

Much ink has been spilled attempting to capture the European Concert's most-important principles, and one could easily cull a list of ten or 20 distinct rules from a handful of Concert sources. Nonetheless, this paper highlights only four such rules—for two reasons. First, I contend that these rules were the most general and foundational and laid the groundwork for smaller, more-context-specific ones. Second, the four I highlight are the principles that have the firmest grounding not only in the best-known secondary historical sources but also in the texts of the treaties themselves (see Table 1).

The first foundational principle of the Vienna System involved designating special status for the most-powerful actors in the system in the first place. Although it is now commonplace to differentiate "great powers" from other states, this differentiation would not have been recognized in Europe prior to the 19th century. But in the post-Napoleonic settlements, and for the first time as a collectivity in history, *(1) the great-power victors of the wars officially granted themselves new status as a separate, more important class of states uniquely fit to dictate the fate of Europe.* Only *they* would be responsible for maintaining peace on the continent and determining what that peace would look like.[18] This principle was first consecrated in a secret clause added to the First Treaty of Paris (May 1814), which noted that "the relations from whence a real and permanent Balance of Power in Europe shall be derived, shall be regulated . . . by the principles determined upon by the Allied Powers amongst themselves."[19] This differentiation was not a mere abstraction born in the heads of statesmen. It also reflected a new material and social reality in which vast military power could be harnessed by the regimes with the largest populations, for the first time in modern history, through the advent of mass conscription.[20]

If the first rule accomplished a necessary preliminary task, the second articulated what was to become the single most important principle at the core of the Concert: *(2) an acknowledgement by the great powers that only together would they establish, defend, and redefine as necessary the political and territorial status quo on the continent.* Simply put, no unilateral territorial changes would be permissible without consent from (or at least consultation with) the great powers acting in concert. This principle was codified across a number of the important agreements and was historically novel in many ways. In the Treaty of Chaumont (March 1814), the four great powers pledged to only negotiate a final peace with France as a collective, single unit. They also committed to this collective for at least 20 years after the war's conclusion should the French revolutionary threat reemerge—a commitment of unprecedented length for *any* state, let alone the most powerful states, to make at the time.[21] The Congress of Vienna produced the Vienna Final Act (June 1815), an agreement that was notable for packaging all of the smaller territorial settlements negotiated at the congress into a single larger treaty. As Jennifer Mitzen has argued, "each individual agreement was given the additional endorsement of being part of the overall plan for continental peace and stability. Through the Final Act, European stability was made indivisible, and it was made the responsibility of all signatories."[22]

Yet the great powers also recognized that *acknowledging* their responsibility to act in concert was only a first step, not the last, in ensuring perpetual stability on the continent. They therefore built into this system *(3) a loose mechanism for consultation and dispute resolution through periodic great-power meetings.*[23] Specifically, the Quadruple Alliance (November 1815) stated,

> To consolidate the connections which at the present moment so closely unite the Four Sovereigns for the happiness of the world, *the High Contracting Parties have agreed to renew their Meetings as fixed periods*, under either the immediate auspices of the Sovereigns themselves, or by their respective Ministers, for the purpose of consulting upon their common interests, and for consideration of the measures which at each of those periods shall be considered the most salutary for the repose and prosperity of Nations and for the maintenance of the Peace of Europe.[24]

If conflicts arose, the great powers would meet and then negotiate among themselves to reach some resolution, resorting to the use of force only as agreed to together and only when necessary to contain a larger disruption to the status quo.[25] At the first of these meetings in 1818, held in Aachen, Prussia,[26] the powers formally expanded their commitment to remain united on matters related to the French threat to *all* European security concerns more generally.[27]

It was at this same 1818 congress that the great powers formally ended the occupation of France, mandated under the Second Treaty of Paris (November 1815). More significantly, they welcomed their former revolutionary adversary—now safely under the rule of the restored monarchy—into their great-power consortium. The now-*five* great powers then jointly declared that "[t]he intimate union established between the Monarchs, who are joint parties to this System, by their own principles, no less than by the interest of their peoples, offers to Europe the most sacred pledge of its future tranquility."[28] This declaration not only reiterated their previous pledge to act only in concert but also indicated the addition of another general principle to the Vienna System: *(4) in assessing polities across Europe that would seek the recognitions of sovereignty and the protections built into the Vienna System, the great powers would henceforth look favorably only upon those with nonrevolutionary and conservative (non-liberal) domestic political institutions.*

This principle is more difficult than the others to pinpoint in the official treaties. Yet careful examination reveals its influence—even as an undercurrent—on many of the era's most-important documents.[29] Concert scholarship most frequently highlights the so-called "Holy Alliance" (November 1815), a brief and vague agreement formalized by Austria, Russia, and Prussia while they were in Paris to negotiate the second peace with France. The document

> **If conflicts arose, the great powers would meet and then negotiate among themselves to reach some resolution, resorting to the use of force only as agreed to together and only when necessary to contain a larger disruption to the status quo.**

makes seemingly little reference to regime type.[30] Yet Concert historians identify its significance not in what it actually says but in the way it was later repurposed by Metternich as a justification for great-power intervention against liberal revolution.[31] Whatever form it took across different documents, this general principle represented an important break from past practice in that it sanctioned great-power interference in the domestic affairs of other European states. As Andreas Osiander explains,

> it was the first attempt in the history of the states system of Europe to provide an abstract criterion for membership of that system—the earlier criterion of Christianity had only been a

necessary, not a sufficient, condition for membership. It was in this capacity that, at Vienna, the concept did have a certain impact: the prominence given to it contributed, perhaps decisively, to the non-re-establishment of earlier non-dynastic actors (Genoa, Venice, Poland). At the same time, it helped to prevent the destruction of another (Saxony).[32]

The episodes Osiander refers to are illustrative: In and after the peace negotiations, the rejection of unilateral conquest (Concert rule 2) was often applied selectively to protect only the autonomy of "traditional" regimes (Saxony, for example), while liberal regimes (Genoa, for instance) were either left to fend for themselves or absorbed by larger autocratic states.[33] When combined with the rule of great-power supremacy, then, this membership principle provided the powers justification for near-constant involvement in the domestic affairs of polities across Europe.[34]

Yet events would soon reveal that not all the powers were as enthusiastic about this principle as others. In hindsight, it appears that its ambiguity in the codified agreements obscured a divide between the two more-liberal western states, Great Britain and France, and the three conservative eastern powers, Austria, Russia, and Prussia—but specifically between Castlereagh and Metternich—over how to interpret a stated preference for traditional regimes and how far to push its implementation. The results of this disagreement are discussed in the next section, which focuses on the behavioral effects of the Concert's principles across the continent.

Why Was the Concert Considered Desirable?

This paper is part of a broader project on contemporary international order. As the project's introductory essay makes clear, use of the term *international order* might imply either (1) an intentionally designed set of institutions or principles (order as input) or (2) the behavioral effects or outcomes of some ordering mechanism (order as output).[35] I have thus far used the term in the input sense but now consider the output (patterns of behavior in 19th-century Europe) that followed from those inputs (the principles of the Vienna System). If this approach has merit, it might be instructive to treat input and output not as alternative conceptions of order but instead as two important components of any single order. After all, we

> In hindsight, it appears that [the principle's] ambiguity in the codified agreements obscured a divide between the two more-liberal western states and the three conservative eastern powers.

care about an order's principles (the inputs) precisely because we believe they often have significant effects on behavior and outcomes (the outputs). This is indeed the case in the instance of the Concert of Europe: Observers today would likely care little about the principles detailed in the previous section if there was not evidence that they influenced European state behavior and continent-wide systemic outcomes in some significant way.

Accordingly, this section identifies four interconnected arenas in which the presence of the Concert's core principles had demonstrable effects on state behavior, interstate outcomes, or both. I contend that the first three of these behavioral outcomes were *self-reinforcing* in the Concert era; that is, the act of states behaving in the patterned ways that constitute those outcomes could only increase the likelihood of those patterns—as well as the overall system—enduring. In contrast, I argue that the Concert's fourth behavioral outcome was *self-destabilizing*; its occurrence and reoccurrence over time actually weakened the perpetuation of the very system that produced it. Both the inputs described in the prior section and the outputs discussed here are depicted in Figure 1.[36]

First, and through the rule of periodic consultation first codified in the Quadruple Alliance, *the Concert system established an unprecedented amount of elite-level contact and consultation through great-power congresses and conferences*. The monumental nature of this endeavor was clear with even the inaugural post-1815 congress at Aix-la-Chapelle—"the first conference ever held between states to regulate international relations in time of peace."[37]

FIGURE 1

Fundamental Principles and Behavioral Effects of the European Concert

Input	Output
Order as great-power agreement on a set of informal rules	**Order as patterned and self-reinforcing behaviors or outcomes**
1. Great powers are designated special status	1. Frequent staging of great-power diplomatic meetings to address crises or settle great-power disputes
2. Great powers pledge to uphold territorial status quo in concert and refrain from unilateral territorial opportunism	a. Avoided open humiliation of great powers
3. Great powers agree to keep their collectivity together through periodic future meetings where necessary	b. Rapid assimilation of new or reconstituted polities into the system
4. *Great powers privilege nonrevolutionary conservative regimes as the most legitimate polities worthy of autonomy and protection (not universal)*	2. Low incidence of armed conflict
	3. High territorial stability
	a. Many instances of individual great-power restraint
	b. Development of great-power spheres-of-influence policing
	c. Successful creation of novel neutral or buffer zones
	4. *Frequent uncertainty and disagreement over the Concert's relationship with revolution and regime type (self-destabilizing)*

The issue of how many of these meetings were subsequently held remains a source of considerable dispute. If one counts only the congresses—those meetings attended by heads of state (e.g., Tsar Alexander) or their principal foreign ministers (e.g., Metternich and Castlereagh)—the number of meetings is less than ten.[38] Including formal conferences, however—those attended by the designated ambassadors to the country hosting the meeting—raises the number of Concert-affiliated meetings considerably.[39] I follow this latter course. (Using the joint congress and conference model adopted by notable historians of the era, such as Charles Webster and F. H. Hinsley, the appendix briefly summarizes the 26 meetings widely recognized as part of the Concert system.[40]) But regardless of the precise number of meetings one counts as part of the Concert, the fact remains that great-power consultation in the 19th century after 1815 was much more frequent than it had ever been before, and it often had an ameliorating effect on state behavior. As Paul Schroeder, the foremost American scholar of the Concert era, has put it, "19th-century statesmen could, with a certain minimum of good will and effort, repeatedly reach viable, agreed-upon outcomes to hotly disputed critical problems. The 18th century simply does not record diplomatic achievements of this kind."[41]

This system of meetings was successful, in part, because it fostered a number of subprinciples and regular practices. I will briefly highlight two. One de facto principle engendered through frequent meetings was the norm that no great power was to be openly disrespected, dishonored, or humiliated. "So long as the European Concert functioned, the five great powers had the assurance that both their legitimate rights and their self-esteem would be respected," writes Richard Elrod. For example, "[u]nder no circumstances did one invite, in any capacity, a state of the second or third rank which was an enemy of a great power."[42] In addition, a regular practice produced through frequent meetings was the rapid and peaceful reintegration of reconstituted polities into the European system of states. The leniency shown to France in both Paris peace settlements and again at the 1818 congress was remarkable. It indicates an understanding that only a relatively satisfied French state could help ensure tranquility on the continent—an unusual and farsighted intuition for diplomats of the time.[43] And this practice did not end with restored France. Even though the unification campaigns that transformed Germany and Italy into powerful states were regarded warily by Concert members in the 1860s and 1870s, it is striking how seamlessly the same great powers nevertheless welcomed these new states into the system.[44]

Second, *the Concert era correlates with a remarkably low incidence of interstate conflict in Europe*. Contrasted with

> One de facto principle engendered through frequent meetings was the norm that no great power was to be openly disrespected, dishonored, or humiliated.

comparable spans of years in both the 18th and 20th centuries, the period between 1815 and 1914—and especially between 1815 and 1853—is notable for significantly lower conflict frequency, conflict duration, and conflict casualties.[45] Most striking was the dearth of armed conflict between the great powers. With the exception of the Crimean War, there were no system-wide conflicts involving all or even most of the great powers between 1815 and 1914, and the 1815–1853 period stands out as particularly peaceful. Yet small states also benefited from peaceful trends and were typically afforded the same protections as the great powers.[46] As Schroeder observes, "[t]here has never been an era in European history before 1815–1848 or since that time when a small state could feel so confident that it would not be the target of conquest or annexation by some great power."[47] Indeed, data show that no European state suffered state death (violent or otherwise) between 1815 and 1859. The few European state deaths occurring in the second half of the century came only as components of the successful German and Italian unification campaigns.[48] None of this peacefulness came about for lack of disagreement or the absence of serious crises in the wake of the Napoleonic Wars. Instead, the accommodating and flexible structure of the Concert system itself afforded the great powers a forum through which to peacefully resolve their disputes.

Given the first two demonstrable effects of the Concert's principles on behavior, the third effect likely comes as little surprise: *an unprecedented degree of territorial stability in Europe.* Simply put, between 1815 and 1853, Europe's political borders changed very little, especially when compared with similar eras in the 18th and 20th

> Between 1815 and 1853, Europe's political borders changed very little, especially when compared with similar eras in the 18th and 20th centuries, as well as the later 19th century.

centuries, as well as the later 19th century. Yet we can also observe three more specific patterns associated with this broader outcome.[49] First, each great power repeatedly restrained itself from territorial opportunism in favor of preserving Concert unity.[50] Instances of the individual great powers practicing restraint are abundant.[51] Russia repeatedly refrained from taking advantage of the declining Ottoman Empire at the urging of its Concert allies, most notably exercising restraint during the Greek Wars of Independence, in the Mehemet Ali crises in Egypt, and in the settlement terms of the numerous Russo-Turkish Wars throughout the century.[52] Great Britain endorsed the neutralization of the Netherlands when it could have easily taken it as a satellite.[53] Even France—the great power most vocal about overturning major elements of the 1815 settlement as the century wore on—often demonstrated remarkable control in the conduct of its foreign policy, especially given its leaders' publicly stated territorial ambitions.[54]

Second, a de facto spheres-of-influence system emerged among the great powers, a development that was a major component of the Troppau and Laibach congresses of 1820–1821.[55] Charles Kupchan elaborates on what this arrangement looked like in practice:

> The power in question did not have a free hand in these spheres, but other members tended to defer to its preferences. Britain oversaw the low countries, Iberian peninsula, and North America, while Russia's sphere extended to parts of eastern Europe, Persia, and Ottoman territory. Austria held sway in northern Italy and jointly managed the German confederation with Prussia. France's reach was initially curtailed after its defeat, but it gradually came to enjoy special influence in the southern and eastern Mediterranean. By recognizing that individual members had particularly salient interests in specific areas, the designation of spheres of influence preempted disputes that might have otherwise jeopardized group cohesion. Such spheres helped manage and contain crises in the periphery by effectively apportioning regional responsibilities among Concert members.[56]

The end result was a system in which each power maintained a sphere, and other Concert members recognized that state's legitimacy to act as it deemed necessary within that sphere.[57] And because the size of these spheres "tracked rather closely [with] their relative military capabilities," each power felt relatively satisfied most of the time with the amount of territory entrusted to its control by the great-power consortium.[58]

That the great powers were particularly successful in setting up buffer zones between them is a third and final component of the Concert's larger territorial stability. In creating political entities to serve as buffers, the architects of the Vienna System sought to strike the right balance between forging polities weak enough to not threaten the powers' security but strong enough to deter the powers from the temptations of unilateral opportunism. The best-known and most-important success story here was the creation of the German Confederation at the Congress of Vienna.[59] Yet the powers achieved similar successes with neutral buffers in the form of the Swiss Confederation, Scandinavia, and the Kingdom of the United Netherlands.[60]

These three patterned outcomes—frequent meetings, low violence, high territorial stability—all fed back into strengthening the Concert system and great-power adherence to its core principles. The fourth did not. Instead, this outcome—*an increasing divide between the powers over whether and how to respond to liberal revolutions across the continent*—likely played at least some part in the Concert's eventual demise. It stemmed, as discussed earlier, from a lack of clarity in the initial agreements regarding the anti-liberal membership principle, as Castlereagh and Metternich each likely believed that his own interpretation had won the day in the 1815 settlements. Yet this only served to temporarily mask what were actually significant differences over the Concert's role in shaping domestic outcomes as much as international ones.

These differences became alarmingly clear once liberal revolutions broke out across multiple European polities in 1820.[61] The eastern powers immediately demanded a congress to organize a concerted response. Yet Great Britain (through Castlereagh) demurred, arguing that unless revolutions clearly threatened international tranquility, they were outside of the Concert's purview. With Castlereagh's absence at the Congresses of Troppau and Laibach (Ljubljana), Metternich led the eastern powers in making sure their own interpretation carried the day. They jointly declared in the Troppau Protocol that

> States which have undergone a change of Government due to revolution, the results of which threaten other states, *ipso facto* cease to be members of the European Alliance, and remain excluded from it until their situation gives guarantees for legal order and stability. If, owing to such situations, immediate danger threatens other states, the Powers bind themselves, by peaceful means, or if be by arms, to bring back the guilty state into the bosom of the Great Alliance.[62]

Castlereagh's quick response to this declaration made clear that Great Britain was not inherently anti-intervention, and in fact supported the right of each great power to take appropriate measures in its own sphere. He emphasized, however, that his country

> cannot admit that this right can receive a general and indiscriminate application to all Revolutionary Movements, without reference to their

> [Troppau] constituted a clear break between the Concert's two foremost architects over one of its foundational principles, a difference of opinion and interpretation between the eastern and western Concert powers that would never be reconciled.

> immediate bearing upon some particular State or States, or be made prospectively the basis of an Alliance. —They regard its exercise as an exception to general principles of the greatest value and importance, and as one that only properly grows out of the circumstances of the special case.[63]

This constituted a clear break between the Concert's two foremost architects over one of its foundational principles, a difference of opinion and interpretation between the eastern and western Concert powers that would never be reconciled.

From this point on, the three eastern powers continuously (though often only sporadically) used the Concert to justify a doctrine of conservative and anti-revolutionary intervention. Likewise, this doctrine was continuously opposed (though not always loudly) by Great Britain and, soon thereafter, France. France's revolution in 1830—coupled with the significant 1832 Reform Act in Britain—installed considerably more-liberal governments in Paris and London. In Great Britain, this included Lord Palmerston, the bombastic and powerful foreign and then prime minister who would almost singlehandedly control British foreign policy for the next 35 years. Palmerston was an unapologetic liberal and an interventionist. Accordingly, he not only continued Castlereagh's tradition of opposing concerted anti-liberal interventions but also began lending rhetorical and even material support to pro-liberal revolutionary causes abroad. These actions, often undertaken with the support of France, unsurprisingly provoked Metternich and the eastern powers to double down on their anti-revolutionary interventionist practices, eroding at least some of the powers' trust in the Concert system and each other.[64] The precise implications of this divide for the long-term prospects of the Vienna System are explored in the next section, which examines the timing and causes of the European Concert's decline.

When and Why Did the Concert Decline?

Scholars continue to disagree about the timing of the Concert's demise. Some argue that it continued functioning all the way until the outbreak of World War I.[65] Support for this perspective comes from the near-continuous series of ambassadorial conferences that persisted in the later 19th and early 20th centuries, right up until 1914. Many of these conferences were successful in resolving particular territorial or colonial issues and allowing the powers to continue acting in concert (see the appendix). Alas, this position is undermined by the failure of even these continuous meetings to stop the kinds of transgressions that had been far less frequent in prior decades. These failures included a new scramble for colonies outside of Europe that nevertheless began infecting continental politics; an inability to control or shape events in the Polish, Italian, and especially German nationalist movements; and, most importantly, a failure to prevent the outbreak of the significant Crimean, Austro-Prussian, and Franco-Prussian great-power conflicts.[66] After a peaceful interlude, major war had returned to Europe.

A second group of scholars argues that the Concert ended much earlier, and actually not long after it had started, with the initial disagreements over liberal revolutions in the early 1820s.[67] Support for this view comes in two forms, the first involving the rapid fall of Castlereagh himself. Overworked, unpopular at home, and possibly mentally ill, the foreign minister committed suicide in the summer of 1822. His successor, George Canning, was a relative Concert skeptic who, early in his tenure, reportedly observed about affairs on the continent that "[t]hings are getting back to a wholesome state again. Every nation for itself and God for us all!"[68] More substantive is the argument that the eastern powers' invocation of the so-called Holy Alliance to suppress liberal revolutions created an irrevocable split that effectively ruined the Concert. Advocates of this interpretation have a point in that the particular issue of domestic revolutions and the Concert was never resolved. Additionally, after 1822, the most-important leaders almost entirely ceased meeting with one another directly (congresses), opting instead to let their intermediaries represent them (conferences) (see the appendix).[69]

Yet it goes too far to suggest that this was simply the end of the Concert. For one thing, there is considerable evidence that the break over liberal revolutions was less a fissure through the *entire* Concert system and more a tacit acknowledgement to agree to disagree on the issues of domestic ideology and intervention from that point on. As Richard Elrod has argued,

> Undeniably, an ideological rift did develop between East and West from the 1820s onward, and the Holy Alliance and nonintervention became convenient symbols and slogans in the resulting debate. Yet concert diplomacy continued to function. It did so because a great-power consensus persisted that transcended political ideology. . . . Despite ideological divergences, the European powers still agreed upon the necessity of peace among themselves and accepted concert diplomacy as the means to manage crises that might jeopardize that peace.[70]

Furthermore, some of the Concert's greatest successes—for example, fostering Russian restraint in the Greek and Turkish Wars, deterring unilateral great-power opportunism against the crumbling Ottoman Empire, and successfully limiting the system-wide effects of additional liberal revolutions in France—came in the years *after* the Troppau and Laibach congresses where this rupture supposedly occurred.

Instead, the strongest perspective on the Concert's demise splits the difference between the others and focuses on two mid-century events: the liberal wave of revolutions across Europe in 1848 and the Crimean War of 1853–1856.[71] The 1848 revolutions seemed almost perfectly designed to fulfill Metternich's worst fears about the possible effects of liberal uprisings. While France had already succumbed to a liberal revolution once before in the Concert era (in 1830), at least a version of the monarchy had once again been restored (albeit a more liberal and limited one). This was not to be in 1848, as France declared itself a republic for the first time since Napoleon's wars, and in fact installed that leader's nephew—the erratic populist, Louis Napoleon—as its first president. The reverberations across Europe were swift and severe. Uprisings in Vienna forced

> The strongest perspective on the Concert's demise focuses on two mid-century events: the liberal wave of revolutions across Europe in 1848 and the Crimean War of 1853–1856.

Metternich's resignation and flight from Austria, an event that galvanized additional liberal successes in Prussia, Italy, Hungary, and elsewhere.[72]

The revolutions mortally wounded the Concert in several ways. They completed the cycle of pushing the last of its architects from power, thus entrusting the Vienna System to leaders who had little prior experience with it or trust in one another. This setback would have been surmountable except for the fact that this new generation of leaders was the first in the Concert era significantly more accountable to—and thus often preoccupied with—their domestic publics. When the Concert was no longer shielded from domestic politics in this way, it became much more difficult for elites to justify to their people their continued cooperation with odious foreign regimes at the expense of seemingly more-immediate national interests. Instances of great powers defecting on the principle of concerted action multiplied. And each time one power pursued unilateral gain at the expense of Concert norms, it became a little easier for others to break out of the Vienna System's virtuous cycle and give in to short-term temptation.

As discussed earlier, this was a process that might have started with the divide over liberal revolutions in the 1820s. In the prior section, I called this divide self-destabilizing because it created a rift that, at least on this issue, continued to grow wider over time. However, whether that rift *alone* would have destroyed the larger Concert even without the 1848 revolutions is an interesting, albeit unanswerable, question. Instead, we know only that the revolutions did happen, rapidly accelerating these destabilizing trends in the process. New leaders not fully socialized to Concert norms plus less insulation from domestic publics unfamiliar with such norms only exacerbated a growing trend of routine violations of norms. Working in combination after 1848, these factors proved to be the Concert's ultimate undoing.

The system's demise would not become as clear to the participants themselves, however, until the outbreak of the Crimean War, a conflict that was more a symptom of the Concert's decline than a cause.[73] For observers today, the war's origins typically appear convoluted and arcane.[74] Yet it is worth noting that they were also convoluted and arcane even to the participants at the time, and this was precisely the kind of conflict—still centered around how

> When the Concert was no longer shielded from domestic politics in this way, it became much more difficult for elites to justify to their people their continued cooperation with odious foreign regimes at the expense of seemingly more-immediate national interests.

to manage the gradual fragmentation of the Ottoman Empire—that an *effective* Concert had done so well to prevent in the decades prior.[75] Instead, an obscure series of disagreements between France and Russia became a full-blown war pitting Russia against Great Britain and France. After three years of brutal and draining warfare, Russia was left utterly defeated, demoralized, and profoundly dissatisfied with the peace settlement imposed upon it. The Russian tsar bears much of the responsibility for this, because it was his rash actions more than any other's that provoked the war. Yet Great Britain and France were also culpable in their dealings with that eastern power that they had so often worked with in concert. For in that treatment, the western powers ultimately "broke the first law of the Concert, 'Thou shalt not challenge or seek to humiliate another great power,' and thereby helped ensure . . . the demise of the Concert itself."[76]

Concert-like conferences continued in the decades after the 1856 peace settlement—and often even succeeded in resolving conflict. Yet the system left in place was no longer capable of forging consensus on those issues that were the most controversial and important to the great powers. A shell of the Vienna System remained—agreement over great-power supremacy and a weakened version of the norm for continuing multilateral meetings, two of the system's four foundational principles. But transnational liberalization had, by this time, rid the system of any conservative solidarity between the Concert's elites, thus destroying one Concert principle.[77] Most importantly, the principle most central to the system's effectiveness—that of settling European political and territorial questions in concert—had clearly withered away by 1856.

What Can We Learn from the Concert?

So, what lessons might American policymakers take from the Concert of Europe as they seek to preserve and enhance their own vision of international order today? Most obviously, they would need to decide if a concert-like arrangement is both desirable and feasible today. While parallels can certainly be drawn between the eras that could put an emulation of the European Concert within the realm of policymakers' imaginations, differences are also abundant. The biggest difference is that today's great-power concert would almost certainly need to be global in order to be balanced and effective. The Concert of Europe, by contrast, was a strictly regional arrangement.[78]

The necessity of a 21st-century concert being a global one also complicates the issue of membership. Whereas membership in the European Concert was fairly obvious, it is a more complicated task to adjudicate which states would merit membership in a global concert today. One recent team of international scholars, the 21st Century Concert Study Group (of which I am a member), has nonetheless attempted to do just that, using criteria to outline configurations of a contemporary global concert with ten, 13, or 17 members.[79] Specifically, the study group identified three criteria for membership: sufficient material power, a demonstrated willingness to participate in global "maintenance," and international recognition by others that this actor's interests must be taken into account on global issues. To quantify these criteria, the group looks to the following eight indicators:

- size of territory
- size of population

- gross national product
- military expenditures
- membership in the Group of Twenty (G20)
- service on the UNSC
- degree of regional centrality
- contributions to recent peace operations mandated and led by the United Nations.

A country that ranks among the top 15 in the world in a particular indicator category is considered to have successfully satisfied that indicator. For the most-selective configuration for a global concert, the ten countries selected are the only states in the world that satisfied at least five of the eight indicators. Those countries are Brazil, the European Union, and India (satisfying all eight indicators); China, Russia, and the United States (satisfying seven of eight indicators); and Indonesia, Japan, Mexico, and Saudi Arabia (satisfying five of eight indicators).[80] Whether or not one agrees with the study group's criteria, its work constitutes an important starting point for the contentious issue of membership in a global 21st-century version of the Concert of Europe.

Regardless of whether American policymakers attempt to explicitly recreate a concert-like arrangement, there remain other important lessons they can take from the European Concert era. First, today's leaders must acknowledge and accept that only an international order built on the existing realities of material power has any chance at stability and longevity. As discussed earlier, the Concert was not constructed as a balance-of-power system. Yet its architects paid careful attention to the *realities* of the European distribution of power in 1815. By designating the great powers as special, the Concert's architects carefully threaded the needle between forging the cooperative bond they believed would be necessary to prevent future chaos and recognizing the collective-action problems inherent in giving *all* of Europe's actors a seat at the table. The particular states recognized as the caretakers of that system were chosen not because they shared particular values or historical ties but because they were the most materially important and capable actors at the end of the Napoleonic Wars. Inequalities of power were also acknowledged even within the great-power consortium, as satisfying Great Britain and Russia took priority over satisfying the lesser great powers, such as Prussia. Likewise, American leaders today might not like the prospect of elevating the positions of any of the emerging BRICS nations (an informal grouping of Brazil, Russia, India, China, and South Africa)—and China in particular—within the existing order. But U.S. leaders must also understand that no truly global order will remain viable for long if these states are not afforded a privileged place in it.

Second, order builders should focus on cultivating loose process norms over particularistic norms about substance. Enactment of the European Concert's first three principles might have been revolutionary at the time, but the principles themselves—great power supremacy, collective preservation of the status quo, and agreements to meet as a collective when necessary—are and were relatively uncontroversial. This is because they are predominantly rules of procedure, not substance. They established only a system through which the great powers could collectively monitor threats to the continent and agree to consult with one another about how to respond to such threats. This said nothing about what kinds of forces constituted a threat, the types of foreign policy behaviors that were inherently legitimate or illegitimate, the internal makeup of the great powers' domestic regimes, or the specific procedures or formal rules that made up the process of consultation. Instead, they established only a forum and focal point through which substantive agreements could be made. And made they were, on such controversial issues as the postwar treatment of France, great-power supremacy in the German states, European relations with the Ottoman Empire, and recognition of new polities across the continent. These achievements came because the procedural order through which they were negotiated had already been put in place. If the powers had attempted to deal with these substantive issues first, they likely would not have achieved such striking success.

This lesson also helps illuminate the weaknesses of both established international institutions and ad hoc, issue-specific forums in contemporary world politics. As recent events have illustrated, assembling ad hoc summits on such issues as the global financial crisis or the Syrian civil war without a prior framework for negotiation is often unlikely to resolve such contentious issues. This is partly because procedural norms for how to approach such negotiations are not in place, while the principal actors often do not have the trust in one another that iterated interaction within even an informal institutional setting builds over time. By focusing on multiple issues rather than being pulled together only for the most contentious and critical ones, states of the Concert of Europe were often able to use side payments on other issues to resolve impasses over the main crises at hand.

> As recent events have illustrated, assembling ad hoc summits on such issues as the global financial crisis or the Syrian civil war without a prior framework for negotiation is often unlikely to resolve such contentious issues.

On the other end of the spectrum, formal institutions—such as the UNSC—are often too burdened by the strict application of cumbersome rules. The unconditional veto of the UNSC's permanent members has often kept the most-contentious issues, but also the most-critical ones, off of its docket throughout its history, both during the Cold War and after. And the rotating nature of the Security Council's other ten seats to short two-year terms consistently prevents the majority of UNSC members from internalizing the organization's procedural norms in the ways that come with iterated interaction over time. The Concert compares favorably to the UNSC in this way, as its looseness and procedural ambiguities were often its saving grace. Had Prussia possessed a formal veto, for

> For American foreign policy today, the lesson here is to establish a clear hierarchy of U.S. priorities both within the international order and within each region of the world.

instance, the issue of Saxony would have upended the Vienna Congress, potentially strangling the Concert in its infancy. Had it survived that episode, it would have become hopelessly deadlocked by 1820 over the issue of anti-liberal intervention. Instead, in both episodes, the Concert's looseness prevailed: The other powers successfully coerced Prussia—which valued consensus above all—into relenting on Saxony in 1815, while the stalemate over liberal revolutions was frequently pushed to the background after 1820 in favor of finding consensus on more-pressing issues.

Third, American order builders must recognize that group consensus is often a valuable commodity in and of itself. The Concert of Europe was most successful when the powers precommitted to working toward consensus on the issue at hand, even with no guarantee about what the substance of that consensus might be. On the other hand, the Concert was least successful when actors came into consultations with a substantive endpoint in mind and then refused to deviate from it. The former approach not only produced a remarkable number of negotiated settlements but also built up deposits of trust and goodwill between the great powers that could be "cashed in" on more-contentious issues later. Conversely, these same powers that, in the Concert's later years, consistently sought short-term gains on issues of the day eventually found that winning these small battles meant losing the larger war. When Austria's Dual Monarchy most needed the Concert in the late 19th and early 20th centuries, for example, its leaders discovered that they had long since spent any deposits of goodwill and, together with similar actions from the other powers, bankrupted the order's very foundations.

For American foreign policy today, the lesson here is to establish a clear hierarchy of U.S. priorities both within the international order and within each region of the world (or at least those under the order's direct purview). In both spheres (international and region-specific), priorities should be differentiated between *primary* and *secondary*, and secondary interests in one sphere should not usurp or take precedence over primary interests in the other. One primary priority in the international order would be to avoid the outright humiliation of another great power in its home region. This priority would be of utmost importance and would be subordinated only if it came into conflict with a primary American interest (internationally or in a particular region). Tensions with China over Taiwan's autonomy could pit primary American priorities in Asia against primary priorities for preserving international order cohesion, in which case primary U.S. priorities should take precedence. Yet in issue areas that American leaders identify as secondary priorities, they should favor avoiding rifts in the order over their own regional interests. U.S. leaders might loathe the idea of getting less than they want over territorial allocation in the South China Sea, for example. Yet they would need to recognize that relaxing their preferences on issues that are not their most vital in a region in order to build consensus and goodwill with others is a necessary price for achieving more-desirable outcomes on higher priorities down the road.

Finally, American order builders should remain wary of domestic political distractions and temptations. I have argued that leaders' increasing preoccupations with public opinion were a principal cause of the European Concert's decay. Although democratically elected leaders cannot afford to disregard public opinion, these leaders must also never become slaves to that opinion if they hope to build and sustain an international order based on great-power consultation and consensus. Instead, leaders must recognize that public opinion will almost always overvalue immediate gains and unilateral demonstrations of resolve and undervalue the utility of compromise and building long-term relationships with foreign elites. This has been a problem for the United States before: Public sentiment has been blamed, for example, for scuttling a Concert-like spheres-of-influence agreement between the great powers—particularly the United States and Soviet Union—in Europe at the end of World War II.[81] American leaders interested in pursuing a concert-like vision for international order today must therefore understand the necessity of either embarking on a sustained public campaign to win over the masses to the policy positions necessary for achieving such a vision or finding a way of shielding the American public and within-concert deliberations from one another.

This, of course, is much easier said than done. Yet solving the problem of domestic politics and public opinion is vital and would have to be at the forefront of considerations for forging a sustainable international order. This is true because leaders giving in to the temptation of domestic pressures can, over time, poison the well of negotiation. Just as Palmerston's strategic use of Russophobia hampered Great Britain's subsequent ability to work with

> Although democratically elected leaders cannot afford to disregard public opinion, these leaders must also never become slaves to that opinion if they hope to build and sustain an international order based on great-power consultation and consensus.

Russia in the mid-1800s, tough talk over such countries as Saudi Arabia, China, and Russia in the 2016 presidential election cycle might have foreclosed future cooperative endeavors that we cannot yet (or ever) even imagine. Admittedly, an American President who appears to collaborate with or kowtow to brutal autocrats will look bad, at least in the moment, from a public relations standpoint. Yet the President must trust that failing to avert chaos or future calamity simply because he or she was unwilling to work with potential competitors will, in the long run, look worse.

After all, Lord Castlereagh was widely reviled by the British public while he was alive, in large part because of his alleged sympathies with the "nefarious" foreign elites with whom he so often collaborated. Onlookers at his funeral procession reportedly hissed at his coffin, and Lord Byron penned the following posthumous "ode" to the long-serving statesman reflecting this widely held sentiment:

> Posterity will ne'er survey
> a Nobler grave than this
> Here lie the bones of Castlereagh:
> Stop, traveller, and piss![82]

Today, Castlereagh is remembered not for these things but for possessing the foresight and prudence to construct a European order that might have averted another great-power war while almost certainly changing the course of international history for the better. The lesson is clear: American policymakers who commit to a strategy of sustained great-power cooperation should understand that they will inevitably suffer some domestic political costs, which is a price tag leaders should be willing to pay for concerted great-power governance.

Taken together, the experiences of the 19th-century European Concert suggest to American policymakers a grand strategy of prudence through diplomatic restraint. That includes restraint in affording rising powers increasing deference commensurate with their increasing material strength; restraint in being content with procedural norms rather than substantive ones as the basis for great-power cooperation; restraint in recognizing that not all American interests are primary priorities, and that subordinating secondary priorities to great-power cohesion is often a worthwhile long-term strategy; and restraint in forsaking short-term political victories at home for the broader objective of effective great-power governance of the international system. Above all, these are the principal lessons of the Concert of Europe.

Appendix. Congresses and Conferences of the Concert Era

Using the joint congress and conference model adopted by notable historians of the era, this appendix briefly summarizes the 26 meetings widely recognized as part of the Concert of Europe system. Congresses—those meetings attended by heads of state or their principal foreign ministers—are distinguished by italics. The remainder are formal conferences—those attended by the designated ambassadors to the country hosting the meeting.[83]

TABLE A.1

The Concert of Europe's Congresses and Conferences

YEAR	CONGRESS OR CONFERENCE	REASON	INVOLVED PARTIES	RESOLUTION
1814–1815	*Vienna*	End of Napoleonic Wars	Great Britain, France, Prussia, Russia, Austria	Vienna Final Act, which combined the 100+ territorial settlements reached throughout the extended conference
1815	*Paris*	Postwar settlement, second defeat of Napoleon	Great Britain, France, Prussia, Russia, Austria	Affirmation of Quadruple Alliance; end of war; resolution on abolishing slave trade
1818	*Aix-la-Chapelle*	French reparations	Great Britain, France, Prussia, Russia, Austria	France accepted war indemnity in return for withdrawal of foreign troops; Quadruple Alliance unofficially added France
1820	*Troppau*	Liberal revolutions in Spain and Naples	Spain, Naples, Sicily, Great Britain, France, Prussia, Russia, Austria	Considered conditions of intervention in Naples, tabled the issue; authorized force against revolutionary states (only Prussia, Russia, and Austria agreed)
1821	*Laibach*	Naples, again	Naples, Sicily, Great Britain, France, Prussia, Russia, Austria	Abolished Neapolitan constitution; Austria then invaded Naples and restored the prior regime
1822	*Verona*	Italian Question (Austrian occupation), Spanish colony revolt, Eastern Question (Turkey in Greece)	Spain, Naples, Sicily, Great Britain, France, Prussia, Russia, Austria	Austria remained in northern Italy; Eastern Question tabled; Russian-led Spanish intervention stopped by Great Britain
1830–1832	London	Belgian independence from the Netherlands	Great Britain, France, Prussia, Russia, Austria, Belgium, Netherlands	Independent Belgian monarchy installed, against the initial wishes of the Holy Alliance; Belgian independence secured by Great Britain and France

TABLE A.1—CONTINUED

YEAR	CONGRESS OR CONFERENCE	REASON	INVOLVED PARTIES	RESOLUTION
1831–1832	Rome	Reform of Papal States in the face of unrest	Great Britain, France, Prussia, Russia, Austria, Papal States	France and Austria jockeyed over Italy; Austria eventually pulled out; reforms failed, although the conflict was contained
1838–1839	London	Belgian independence, again	Great Britain, France, Prussia, Russia, Austria, Belgium	Reopening of the 1832 conference; Belgium acquiesced to British, Prussian, and Austrian demands
1839	Vienna	Internal Ottoman conflict between Sultan and Egyptian vassal (Eastern Question)	Great Britain, France, Prussia, Russia, Austria, Turkey	Metternich proposed a Viennese meeting, although a diplomatic issue involving Russia caused it to fall apart
1840–1841	London	Continued internal Ottoman conflict (Eastern Question)	Great Britain, France, Prussia, Russia, Austria, Turkey	Russia had a change of heart from the previous meeting; the powers signed the second London Straits Convention, which closed the Bosporus and Dardanelles to warships, including ships of Turkey's allies and enemies
1850–1852	London	Schleswig-Holstein War	Great Britain, France, Prussia, Russia, Austria, Denmark, Sweden	Restoration of the prewar status quo: The London Protocol, which stated that Schleswig and Holstein were part of Denmark, was signed in 1850 and revised in 1852
1853	Vienna	Crimean War outbreak (Eastern Question)	Great Britain, France, Prussia, Russia, Austria, Turkey	Produced the Vienna Note, a compromise to end the conflict between Russia and Turkey, but the agreement was ultimately toothless
1855	Vienna	Proposal to end Crimean War (Eastern Question)	Great Britain, France, Prussia, Russia, Austria, Turkey, Sweden	Diplomatic pressure by Great Britain and France successfully persuaded Russia to officially attend the upcoming Congress of Paris
1856	*Paris*	Resolution of Crimean War (Eastern Question)	Great Britain, France, Turkey, Russia	Territories of Russia and Turkey restored to prewar boundaries; neutralization of the Black Sea; external guarantee of Turkey's independence
1860–1861	Paris	Syria peasant uprising	Great Britain, France, Prussia, Russia, Austria, Turkey	Multilateral great power intervention and peacekeeping operation
1864	London	Schleswig-Holstein conflict, again	Great Britain, France, Prussia, Russia, Austria, Denmark	Ceasefire failed when Prussia and Austria opposed any negotiated settlement to the issue and ultimately invaded Denmark

TABLE A.1—CONTINUED

YEAR	CONGRESS OR CONFERENCE	REASON	INVOLVED PARTIES	RESOLUTION
1867	London	Luxembourg crisis	Great Britain, France, Prussia, Russia, Austria, Luxembourg, Netherlands	Prussia occupied Luxembourg, but the Netherlands claimed it under 1815 accords; peace was eventually preserved through the great powers securing Luxembourg's independence and neutrality
1869	Paris	Cretan revolt	Great Britain, France, Prussia, Russia, Austria, Turkey, United States	Resolution of Cretan rebellion from Turkey: the United States declined Crete's plea for assistance; status quo restored and Cretans suppressed
1871	London	Black Sea militarization	Great Britain, France, Prussia, Russia, Austria, Italy	Overturned neutralization of the Black Sea
1876	Constantinople	Bosnian reforms (Eastern Question)	Great Britain, France, Germany, Russia, Austria, Italy	Addressed reforms in Bosnia/Bulgarian Ottoman territory deemed necessary after the Herzegovinian Uprising the year before; created autonomous Bosnian province and two Bulgarian provinces
1878	*Berlin*	Resolution of Russo-Turkish War	Great Britain, France, Germany, Russia, Austria, Turkey	Revised the peace settlement of the Russo-Turkish War; allowed Austria to occupy Bosnia and Herzegovina
1880	Madrid	Independence of Morocco	Great Britain, France, Germany, Russia, Austria, Spain, Denmark, Morocco, Portugal, Sweden, Norway, Italy, Belgium, United States	Morocco's "independence" guaranteed by France and allies after the war with Spain (1859) is concluded; these allies subsequently controlled Morocco's banking, trade, police, etc.
1884–1885	Berlin	Scramble for African colonies	Great Britain, France, Germany, Russia, Austria	Legitimized prior European "effective occupation" of colonies in Africa; established free trade throughout the Congo Basin; opened the Niger and Congo rivers to international traffic; nominally declared the sovereignty of the Congo Free State, a polity that nonetheless remained under Belgium's control
1906	Algeçiras	"First" Moroccan Crisis	Great Britain, France, Germany, Russia, Austria, Morocco	Germany sought a truly independent Morocco; France and others opposed; Morocco independence was nominally declared, but France ultimately won and gave up little of its control in Morocco
1912–1913	London	Balkans	Great Britain, France, Prussia, Russia, Austria, Serbia	Ended first Balkan War but sowed seeds for future crisis

Notes

[1] Henry A. Kissinger, *A World Restored: Metternich, Castlereagh and the Problems of Peace, 1812–1822*, Boston: Houghton Mifflin Company, 1957, p. 5.

[2] See, for example, Henry A. Kissinger, *Diplomacy*, New York: Simon and Schuster, 1994; and Henry A. Kissinger, *World Order*, New York: Penguin, 2014.

[3] Mark Mazower, *Governing the World: The History of an Idea*, New York: The Penguin Press, 2012, pp. 123–125, 195–197.

[4] Mazower, 2012, p. 11.

[5] See, for instance, Philip Zelikow, "The New Concert of Europe," *Survival*, Vol. 34, No. 2, 1992; Richard Rosecrance, "A New Concert of Powers," *Foreign Affairs*, Vol. 71, No. 2, Spring 1992; Charles A. Kupchan and Adam Mount, "The Autonomy Rule," *Democracy*, Vol. 12, 2009; 21st Century Concert Study Group, *A Twenty-First Century Concert of Powers—Promoting Great Power Multilateralism for the Post-Transatlantic Era*, Frankfurt: Peace Research Institute Frankfurt, 2014; and Thomas Wright, "The Rise and Fall of the Unipolar Concert," *Washington Quarterly*, Vol. 37, No. 4, 2015.

[6] See John J. Mearsheimer, "The False Promise of International Institutions," *International Security*, Vol. 19, No. 3, 1994/1995, pp. 34–37; Korina Kagan, "The Myth of the European Concert: The Realist-Institutionalist Debate and Great Power Behavior in the Eastern Question, 1821–41," *Security Studies*, Vol. 7, No. 2, 1997; Matthew Rendall, "Russia, the Concert of Europe, and Greece 1821–29: A Test of Hypotheses About the Vienna System," *Security Studies*, Vol. 9, No. 4, Summer 2000; and Branislav L. Slantchev, "Territory and Commitment: The Concert of Europe as Self-Enforcing Equilibrium," *Security Studies*, Vol. 14, No. 4, 2005.

[7] These departures are discussed in the later section on why the Concert is seen as desirable.

[8] Jennifer Mitzen, *Power in Concert: The Nineteenth-Century Origins of Global Governance*, Chicago: University of Chicago Press, 2013, pp. 22–23.

[9] See, for example, Richard B. Elrod, "The Concert of Europe: A Fresh Look at the International System," *World Politics*, Vol. 28, No. 2, 1976, pp. 161–162; Paul W. Schroeder, "The 19th-Century International System: Changes in the Structure," *World Politics*, Vol. 39, No. 1, 1986; Paul W. Schroeder, "Did the Vienna Settlement Rest on a Balance of Power?" *American Historical Review*, Vol. 97, No. 2, 1992; and Slantchev, 2005, pp. 577–580.

[10] On the former characterization, see Robert Jervis, "Security Regimes," in Stephen D. Krasner, ed., *International Regimes*, Ithaca, N.Y.: Cornell University Press, 1983; Robert Jervis, "From Balance to Concert: A Study of International Security Cooperation," *World Politics*, Vol. 38, No. 1, 1985; Charles A. Kupchan and Clifford A. Kupchan, "Concerts, Collective Security, and the Future of Europe," *International Security*, Vol. 16, No. 1, 1991; and Dan Lindley, "Avoiding Tragedy in Power Politics: The Concert of Europe, Transparency, and Crisis Management," *Security Studies*, Vol. 13, No. 2, 2003. On the latter characterization, see G. John Ikenberry, *After Victory: Institutions, Strategic Restraint, and the Rebuilding of Order After Major Wars*, Princeton, N.J.: Princeton University Press, 2001, Chapter 4.

[11] On developing collective intentions and interests, see Charles A. Kupchan, *How Enemies Become Friends: The Sources of Stable Peace*, Princeton, N.J.: Princeton University Press, 2010, Chapter 5; and Mitzen, 2013. On developing a common transnational identity, see Bruce Cronin, *Community Under Anarchy: Transnational Identity and the Evolution of Cooperation*, New York: Columbia University Press, 1999, Chapter 3; Mark L. Haas, *The Ideological Origins of Great Power Politics, 1789–1989*, Ithaca, N.Y.: Cornell University Press, 2005, Chapter 3; and Paul W. Schroeder, *The Transformation of European Politics, 1763–1848*, New York: Oxford University Press, 1994.

[12] Elrod, 1976; Slantchev, 2005, especially pp. 591, 606; and Kupchan, 2010, pp. 196–200.

[13] Lindley, 2003.

[14] The Concert case studies in Cronin (1999), Haas (2005), and Kupchan (2010, pp. 200, 238–239) all endorse elements of this argument.

[15] In existing scholarship on the Concert, this definition is probably most similar to Elrod, 1976.

[16] On the basis of such fears—both in the French Revolutionary instance and in ideological conflicts and contests more generally—see John M. Owen IV, *The Clash of Ideas in World Politics: Transnational Networks, States, and Regime Change, 1510–2010*, Princeton, N.J.: Princeton University Press, 2010.

[17] For an excellent recent summary of the Vienna System and its significance, see Mitzen, 2013, Chapter 3.

[18] See, for example, Elrod, 1976, pp. 163–165; and David King, *Vienna, 1814: How the Conquerors of Napoleon Made Love, War, and Peace at the Congress of Vienna*, New York: Three Rivers Press, 2008, p. 50.

[19] Edward Hertslet. *The Map of Europe by Treaty: Political and Territorial Changes Since the General Peace of 1814*, Vol. 1, London: Butterworths, 1875, p. 18.

[20] See Andreas Osiander, *The States System of Europe, 1640–1990: Peacemaking and the Conditions of International Stability*, New York: Oxford University Press, 1994, p. 236. A large population was, of course, a necessary but not always sufficient condition for successful mass conscription. Bureaucratic organization (on the part of the state) and a unifying nationalist spirit (on the part of the masses) were often necessary preconditions for transforming a large population into a large and effective military.

[21] Ikenberry, 2001, pp. 93–96; King, 2008, p. 247.

[22] Mitzen, 2013, p. 95.

[23] Kissinger, 1957, p. 186; King, 2008, pp. 309–310.

[24] Hertslet, 1875, p. 375, emphasis added.

[25] See Elrod, 1976, pp. 163–168; Schroeder, 1992.

[26] Aachen was known as Aix-la-Chapelle when it was previously held by France and is often still referred to by this name in official documents (most likely because the treaties were originally written in French).

[27] Specifically, "The five Powers . . are firmly resolved never to depart, neither in their mutual Relations, nor in those which bind them to other states, from principles of intimate union which hitherto presided over all their common relations and interests" (quoted in Cronin, 1999, pp. 60–61). Although this seemingly high-minded rhetoric might sound like public relations gibberish, it was actually part of a secret protocol, just for the great powers, and was not released to the public or other states.

[28] Hertslet, 1875, p. 573; see also F. R. Bridge and Roger Bullen, *The Great Powers and the European States System, 1814–1914*, 2nd ed., Harlow, U.K.: Pearson Education Limited, 2005, pp. 4–11.

[29] In the Second Treaty of Paris, for instance, the powers praised themselves for preserving "France and Europe from the convulsions with which they were menaced by the late enterprise of Napoleon Bonaparte, and by the Revolutionary system reproduced in France, to promote its success." They also noted how "the order of things . . . had been happily re-established in France" with the restoration of the Bourbon monarchy before going on to articulate a similar goal of "restoring between France and her Neighbours those relations of reciprocal confidence and goodwill which the fatal effects of the Revolution and of the system of conquest had for so long a time disturbed." (Hertslet, 1875, pp. 342–343)

[30] The document states, "[T]he Three contracting Monarchs will remain united by the bonds of a true and indissoluble fraternity, . . . [and] will, on all occasions and in all places, lend each other aid and assistance" (Hertslet, 1875, p. 318).

[31] Among others, Kissinger argues that Alexander's initial idea for and articulation of the Holy Alliance were unintelligible; it was Metternich who later gave the pact tangible meaning by connecting it to liberal regimes and domestic disturbances. See Kissinger, 1957, p. 189; 1994, p. 83.

[32] Osiander, 1994, p. 223.

[33] King, 2008, p. 317.

[34] See Martha Finnemore, *The Purpose of Intervention: Changing Beliefs About the Use of Force*, Ithaca, N.Y.: Cornell University Press, 2003, pp. 117–121.

[35] Michael J. Mazarr, Miranda Priebe, Andrew Radin, and Astrid Cevallos, *Understanding the Current International Order*, Santa Monica, Calif.: RAND Corporation, RR-1598-OSD, 2016, pp. 9–10.

[36] Figure 1 intentionally mimics the design of Figure 2.1 in Mazarr et al., 2016, p. 9.

[37] Ikenberry, 2001, p. 81.

[38] This is the view taken in Kissinger, 1957, for instance.

[39] On this distinction, see F. H. Hinsley, *Power and the Pursuit of Peace: Theory and Practice in the History of Relations Between States*, New York: Cambridge University Press, 1963, pp. 213–214.

[40] Charles Webster, *The Art and Practice of Diplomacy*, New York: Barnes & Noble, 1961, p. 69; and Hinsley, 1963, p. 214. For a view that identifies a considerably greater number of important ambassadorial conferences, see Matthias Schulz, *Normen und Praxis: Das Europäische Konzert der Grossmächte als Sicherheitsrat, 1815–1860*, Munich: Oldenbourg, 2009, pp. 684–685.

[41] Schroeder, 1986, p. 3.

[42] Elrod, 1976, pp. 166–167. See also Cronin, 1999, pp. 63–64.

[43] Castlereagh in particular succeeded in convincing others at the Vienna Congress that "if we push things now to an extremity, we leave the [newly reinstalled French] king no resource in the eyes of his own people but to disavow us; and, once committed against us in sentiment," European stability would be lost (Osiander, 1994, p. 202). On Castlereagh's role in this smooth reintegration, see Kissinger, 1957, pp. 177–180. That France itself was not blamed nor condemned for Napoleon's escape and resumption of war in 1815 is also remarkable, as it was probably "the first time in history that states had effectively declared war on a single person" rather than the larger polity that person commanded (King, 2008, p. 242).

[44] See Schroeder, 1986, pp. 5–9.

[45] Matthias Schulz, "Did Norms Matter in Nineteenth-Century International Relations? Progress and Decline in the 'Culture of Peace' Before World War I," in Holger Afflerbach and David Stevenson, eds., *An Improbable War: The Outbreak of World War I and European Culture before 1914*, New York: Berghan Books, 2007. Schulz and many others cite Jack Levy's comprehensive war data to support this point. See Jack Levy, *War in the Modern Great Power System, 1495–1975*, Lexington, Ky.: University of Kentucky Press, 1983, Chapter 4.

[46] Elrod, 1976, pp. 165–166.

[47] Schroeder, 1986, p. 25.

[48] See the state death data collected for and presented in Tanisha M. Fazal, *State Death: The Politics and Geography of Conquest, Occupation, and Annexation*, Princeton, N.J.: Princeton University Press, 2007, pp. 21–23.

[49] I leave to the reader whether these three patterns are best characterized as indicators/effects of territorial stability or causes/mechanisms for it. I suspect it is some combination of both.

[50] In an agreement recognizing the independence of Greece from the Ottoman Empire in the 1820s, for instance, the great powers jointly declared that "[t]he Contracting Powers will not seek, in these arrangements, any augmentation of territory, any exclusive influence, or any commercial advantage for their Subjects, which those of every other Nation may not equally obtain" (Jervis, 1985, p. 73).

[51] On this point, and the relative frequency of each great power's instances of unilateral Concert violations more generally, see Schulz, 2007, pp. 52–54.

[52] Rendall, 2000; Haas, 2005, pp. 75–90.

[53] Kupchan, 2010, pp. 190–192.

[54] Explaining why he was ceasing French opportunism in Egypt in 1841, for instance, King Louis Philippe argued that France "wishes to maintain the European equilibrium, the care of which is the responsibility of all the Great Powers. Its preservation must be their glory and their main ambition" (Kupchan, 2010, p. 192).

[55] Slantchev, 2005, p. 585.

[56] Kupchan, 2010, pp. 192–193.

[57] This is the principal reason I do not include "nonintervention" as a general, foundational rule of the Concert system, as some others do. Great powers *were* allowed to unilaterally intervene in polities' affairs within their respective spheres.

[58] Slantchev, 2005, p. 590. Slantchev additionally stresses the importance of the two most-powerful states, Great Britain and Russia, having been the most satisfied with their spheres in the wake of the settlements.

[59] Kissinger aptly summarizes the Congress of Vienna's ingenious solution to the German question: "The German Confederation was too divided to take offensive action yet cohesive enough to resist foreign invasions into its territory. This arrangement provided an obstacle to the invasion of Central Europe without constituting a threat to the two major powers on its flanks, Russia to the east and France to the west" (Kissinger, 2014, p. 64).

[60] Schroeder, 1986, pp. 17–20.

[61] On these revolutions and reactions, see Owen, 2010, pp. 147–148.

[62] René Albrecht-Carrié, *The Concert of Europe*, New York: Walker & Company, 1968, p. 48. This statement was the most notable part of the preliminary protocols of the agreements at the Congress of Troppau and is often now known simply as the "Troppau Protocol."

[63] See document 5 in Albrecht-Carrié, 1968, pp. 49–52.

[64] The clearest example of this erosion is the 1833 Münchengrätz Convention. On the development of these dueling ideological camps within the concert, see Haas, 2005, Chapter 3.

[65] Perspectives that fit in this camp include René Albrecht-Carrié's classic study; Matthias Schulz's writings in English and German; Richard Langhorne's various works; R. J. Crampton, "The Decline of the Concert of Europe in the Balkans, 1913–1914," *Slavonic and East European Review*, Vol. 52, 1974; and Georges-Henri Soutou, "Was There a European Order in the Twentieth Century? From the Concert of Europe to the End of the Cold War," *Contemporary European History*, Vol. 9, No. 3, 2000.

[66] These examples are taken from Schulz, 2007, pp. 49–50. That these failures are admitted by Schulz, one of the strongest proponents of the position that the Concert continued until 1914, only strengthens the argument that they were profound shortcomings.

[67] Notable scholars in this camp include Henry Kissinger and Dan Lindley, as well as classic works like Harold George Nicolson, *The Congress of Vienna: A Study in Allied Unity: 1812–1822*, London: Constable & Co ltd., 1946; and Carsten Holbraad, *The Concert of Europe: A Study in German and British International Theory*, New York: Barnes & Noble, 1970.

[68] Jervis, 1983, p. 180.

[69] The exceptions came only in 1856 and 1878, at the conclusion of two particularly important wars (the Crimean War and the Russo-Turkish War, respectively).

[70] Elrod, 1976, pp. 171–172. As evidence for this interpretation, Castlereagh even closed his instructions to Britain's ambassadors in reaction to the Troppau Protocol with the following optimistic statement: "The difference of sentiment which prevails between them and the Court of London on this matter, you may declare, can make no alteration whatever in the cordiality and harmony of the Alliance on any other subject, or abate their common zeal in giving the most complete effect to all their existing engagements" (Albrecht-Carrié, 1968, pp. 51–52).

[71] Notable scholars in this camp include Bruce Cronin, Richard Elrod, Mark Haas, Charles Kupchan, Paul Schroeder, and Branislav Slantchev.

[72] Compelling narratives of these events include Charles Pouthas, "The Revolutions of 1848," in J. P. T. Bury, ed., *The New Cambridge Modern History*, Vol. 10, *The Zenith of European Power 1830–70*, New York: Cambridge University Press, 1960; E. J. Hobsbawm, *The Age of Revolution, 1789–1848*, New York: The World Publishing Company, 1962; and Mike Rapport, *1848: Year of Revolution*, New York: Basic Books, 2010.

[73] While most accounts that mark the Concert's end at mid-century treat the Crimean War in this way, some see the great-power decisions for that war as contingent and treat the war itself as the causal force ending the Vienna System. See, for instance, Mitzen, 2013, Chapter 6.

[74] That said, good starting places include Norman Rich, *Why the Crimean War? A Cautionary Tale*, Hanover, N.H.: University Press of New England, 1985; David Wetzel, *The Crimean War: A Diplomatic History*, New York: Columbia University Press, 1985; and Orlando Figes, *The Crimean War: A History*, New York: Metropolitan Books, 2011.

[75] Kupchan, 2010, pp. 239–240.

[76] Paul W. Schroeder, *Austria, Great Britain, and the Crimean War: The Destruction of the European Concert*, Ithaca, N.Y.: Cornell University Press, 1972, p. 409.

[77] Kissinger in particular believes that this was an important bond, especially in restraining competition between Russia, Prussia, and Austria in the east. See, for instance, Kissinger, 1994, p. 94.

[78] According to some observers, limiting its purview to the European continent was, in fact, one of the sources of the original Concert's success. See, for instance, Schroeder, 1986.

[79] 21st Century Concert Study Group, 2014.

[80] See 21st Century Concert Study Group, 2014, pp. 36–44.

[81] See, for instance, Marc Trachtenberg, *A Constructed Peace: The Making of the European Settlement, 1945–1963*, Princeton, N.J.: Princeton University Press, 1999, Chapters 1–2; and Colin Dueck, *Reluctant Crusaders: Power, Culture, and Change in American Grand Strategy*, Princeton, N.J.: Princeton University Press, 2008, Chapter 4.

[82] Lord Byron, "Epitaph for Lord Castlereagh," in *The Complete Works of Lord Byron*, Paris: Baudry's European Library, 1835.

[83] I thank Drake MacFarlane for excellent research assistance in compiling the data for this appendix. The sources for the appendix are Albrecht-Carrié, 1968; Bridge and Bullen, 2005; Kalevi J. Holsti, *Peace and War: Armed Conflicts and International Order 1648–1989*, Cambridge, U.K.: Cambridge University Press, 1991; A. J. P. Taylor, *The Struggle for Mastery in Europe 1848–1918*, 1st ed., Oxford, U.K.: Oxford University Press, 1954; and Webster, 1961.

Bibliography

21st Century Concert Study Group, *A Twenty-First Century Concert of Powers—Promoting Great Power Multilateralism for the Post-Transatlantic Era*, Frankfurt: Peace Research Institute Frankfurt, 2014.

Albrecht-Carrié, René, *The Concert of Europe*, New York: Walker & Company, 1968.

Bridge, F. R., and Roger Bullen, *The Great Powers and the European States System 1814–1914*, 2nd ed., Harlow, U.K.: Pearson Education Limited, 2005.

Craig, Gordon A., *Europe, 1815–1914*, 2nd ed., New York: Holt, Rinehart, and Winston, 1966.

Crampton, R. J., "The Decline of the Concert of Europe in the Balkans, 1913–1914," *Slavonic and East European Review*, Vol. 52, 1974, pp. 393–419.

Cronin, Bruce, *Community Under Anarchy: Transnational Identity and the Evolution of Cooperation*, New York: Columbia University Press, 1999.

Dueck, Colin, *Reluctant Crusaders: Power, Culture, and Change in American Grand Strategy*, Princeton, N.J.: Princeton University Press, 2008.

Elrod, Richard B., "The Concert of Europe: A Fresh Look at the International System," *World Politics*, Vol. 28, No. 2, 1976, pp. 159–174.

Fazal, Tanisha M., *State Death: The Politics and Geography of Conquest, Occupation, and Annexation*, Princeton, N.J.: Princeton University Press, 2007.

Figes, Orlando, *The Crimean War: A History*, New York: Metropolitan Books, 2011.

Finnemore, Martha, *The Purpose of Intervention: Changing Beliefs about the Use of Force*, Ithaca, N.Y.: Cornell University Press, 2003.

Haas, Mark L., *The Ideological Origins of Great Power Politics, 1789–1989*, Ithaca, N.Y.: Cornell University Press, 2005.

Hertslet, Edward, *The Map of Europe by Treaty: Political and Territorial Changes Since the General Peace of 1814*, Vol. 1, London: Butterworths, 1875.

Hinsley, F. H., *Power and the Pursuit of Peace: Theory and Practice in the History of Relations Between States*, New York: Cambridge University Press, 1963.

Hobsbawm, E. J., *The Age of Revolution, 1789–1848*, New York: The World Publishing Company, 1962.

Holbraad, Carsten, *The Concert of Europe: A Study in German and British International Theory*, New York: Barnes & Noble, 1970.

Holsti, Kalevi J., *Peace and War: Armed Conflicts and International Order, 1648–1989*, Cambridge, U.K.: Cambridge University Press, 1991.

Ikenberry, G. John, *After Victory: Institutions, Strategic Restraint, and the Rebuilding of Order After Major Wars*, Princeton, N.J.: Princeton University Press, 2001.

Jervis, Robert, "Security Regimes," in Stephen D. Krasner, ed., *International Regimes*, Ithaca, N.Y.: Cornell University Press, 1983, pp. 173-194.

———, "From Balance to Concert: A Study of International Security Cooperation," *World Politics*, Vol. 38, No. 1, 1985, pp. 58–79.

Kagan, Korina, "The Myth of the European Concert: The Realist-Institutionalist Debate and Great Power Behavior in the Eastern Question, 1821–41," *Security Studies*, Vol. 7, No. 2, 1997, pp. 1–57.

King, David, *Vienna, 1814: How the Conquerors of Napoleon Made Love, War, and Peace at the Congress of Vienna*, New York: Three Rivers Press, 2008.

Kissinger, Henry A., *A World Restored: Metternich, Castlereagh and the Problems of Peace, 1812–1822*, Boston: Houghton Mifflin Company, 1957.

———, *Diplomacy*, New York: Simon and Schuster, 1994.

———, *World Order*, New York: Penguin, 2014.

Kupchan, Charles A., *How Enemies Become Friends: The Sources of Stable Peace*, Princeton, N.J.: Princeton University Press, 2010.

Kupchan, Charles A., and Clifford A. Kupchan, "Concerts, Collective Security, and the Future of Europe," *International Security*, Vol. 16, No. 1, 1991, pp. 114–161.

Kupchan, Charles A., and Clifford A. Kupchan, "The Promise of Collective Security," *International Security*, Vol. 20, No. 1, 1995, pp. 52–61.

Kupchan, Charles A., and Adam Mount, "The Autonomy Rule," *Democracy*, Vol. 12, 2009.

Langhorne, Richard, "Reflections on the Significance of the Congress of Vienna," *Review of International Studies*, Vol. 12, No. 4, 1986, pp. 313–324.

———, "Establishing International Organisations: The Concert and the League," *Diplomacy & Statecraft*, Vol. 1, No. 1, 1990, pp. 1–18.

Levy, Jack, *War in the Modern Great Power System, 1475–1975*, Lexington, Ky.: The University Press of Kentucky, 1983.

Lindley, Dan, "Avoiding Tragedy in Power Politics: The Concert of Europe, Transparency, and Crisis Management," *Security Studies*, Vol. 13, No. 2, 2003, pp. 195–229.

Lord Byron, "Epitaph for Lord Castlereagh," in *The Complete Works of Lord Byron*, Paris: Baudry's European Library, 1835.

May, Arthur, *The Age of Metternich: 1814–1848*, New York: Henry Holt and Company, 1933.

Mazarr, Michael J., Miranda Priebe, Andrew Radin, and Astrid Cevallos, *Understanding the Current International Order*, Santa Monica, Calif.: RAND Corporation, RR-1598-OSD, 2016. As of November 15, 2016:
http://www.rand.org/pubs/research_reports/RR1598.html

Mazower, Mark, *Governing the World: The History of an Idea*, New York: The Penguin Press, 2012.

Mearsheimer, John J., "The False Promise of International Institutions," *International Security*, Vol. 19, No. 3, 1994/1995, pp. 5–49.

Mitzen, Jennifer, *Power in Concert: The Nineteenth-Century Origins of Global Governance*, Chicago: University of Chicago Press, 2013.

Nicolson, Harold George, *The Congress of Vienna: A Study in Allied Unity: 1812–1822*, London: Constable & Co ltd., 1946.

Osiander, Andreas, *The States System of Europe, 1640–1990: Peacemaking and the Conditions of International Stability*, New York: Oxford University Press, 1994.

Owen, John M. IV, *The Clash of Ideas in World Politics: Transnational Networks, States, and Regime Change, 1510–2010*, Princeton, N.J.: Princeton University Press, 2010.

Pouthas, Charles, "The Revolutions of 1848," in J. P. T. Bury, ed., *The New Cambridge Modern History*, Vol. 10, *The Zenith of European Power 1830–70*, New York: Cambridge University Press, 1960.

Rapport, Mike, *1848: Year of Revolution*, New York: Basic Books, 2008.

Rendall, Matthew, "Russia, the Concert of Europe, and Greece, 1821–29: A Test of Hypotheses About the Vienna System," *Security Studies*, Vol. 9, No. 4, 2000, pp. 52–90.

———, "A Qualified Success for Collective Security: The Concert of Europe and the Belgian Crisis, 1831," *Diplomacy and Statecraft*, Vol. 18, 2007, pp. 271–295.

Rich, Norman, *Why the Crimean War? A Cautionary Tale*, Hanover, N.H.: University Press of New England, 1985.

Rosecrance, Richard, "A New Concert of Powers," *Foreign Affairs*, Vol. 71, No. 2, Spring 1992.

Schroeder, Paul W., *Austria, Great Britain, and the Crimean War: The Destruction of the European Concert*, Ithaca, N.Y.: Cornell University Press, 1972.

———, "The 19th-Century International System: Changes in the Structure," *World Politics*, Vol. 39, No. 1, 1986, 1–26.

———, "Did the Vienna Settlement Rest on a Balance of Power?" *American Historical Review*, Vol. 97, No. 3, 1992, pp. 683–706.

———, *The Transformation of European Politics, 1763–1848*, Oxford, U.K.: Oxford University Press, 1994.

Schulz, Matthias, "Did Norms Matter in Nineteenth-Century International Relations? Progress and Decline in the 'Culture of Peace' Before World War I," in Holger Afflerbach and David Stevenson, eds., *An Improbable War: The Outbreak of World War I and European Culture Before 1914*, New York: Berghan Books, 2007, pp. 43–59.

———, *Normen und Praxis: Das Europäische Konzert der Grossmächte als Sicherheitstrat, 1815–1860*, Munich: Oldenbourg, 2009.

Sked, Alan, ed., *Europe's Balance of Power, 1815–1848*, London: Macmillan Press, 1979.

Slantchev, Branislav L., "Territory and Commitment: The Concert of Europe as Self-Enforcing Equilibrium," *Security Studies*, Vol. 14, No. 4, 2005, pp. 565–606.

Soutou, Georges-Henri, "Was There a European Order in the Twentieth Century? From the Concert of Europe to the End of the Cold War," *Contemporary European History*, Vol. 9, No. 3, 2000, pp. 329–353.

Taylor, A. J. P., *The Struggle for Mastery in Europe 1848–1918*, 1st ed., Oxford, U.K.: Oxford University Press, 1954.

Trachtenberg, Marc, *A Constructed Peace: The Making of the European Settlement, 1945–1963*, Princeton, N.J.: Princeton University Press, 1999.

Webster, Charles, *The Foreign Policy of Castlereagh*, London: G. Bell and Sons, 1931.

———, *The Art and Practice of Diplomacy*, New York: Barnes & Noble, 1961.

Wetzel, David, *The Crimean War: A Diplomatic History*, New York: Columbia University Press, 1985.

Wright, Thomas, "The Rise and Fall of the Unipolar Concert," *Washington Quarterly*, Vol. 37, No. 4, 2015.

Zelikow, Philip, "The New Concert of Europe," *Survival*, Vol. 34, No. 2, 1992

About the Author

Kyle Lascurettes is assistant professor of international affairs at Lewis & Clark College in Portland, Oregon. He specializes in international relations theory, international order, and the intersection of global governance and international security. He is currently finishing a book about great powers and world order throughout the modern international system. He holds a Ph.D. in foreign affairs from the University of Virginia.